Vincent Works William Sugg

Modern Street Lighting

Vincent Works William Sugg

Modern Street Lighting

ISBN/EAN: 9783337254056

Printed in Europe, USA, Canada, Australia, Japan

Cover: Foto ©berggeist007 / pixelio.de

More available books at **www.hansebooks.com**

MODERN
STREET LIGHTING.

BY

WILLIAM SUGG,

ASSOCIATE INST. CIV. ENGINEERS.

LONDON:

VINCENT WORKS, VINCENT STREET,

WESTMINSTER.

1871.

LONDON:
PRINTED BY BLANCHARD AND SONS, 62, MILLBANK STREET,
AND 6 & 7, HORSE AND GROOM YARD, WESTMINSTER.

PREFACE.

This little Work is issued by the Author in order to supply a want which, he has abundant proof, has been felt by many who have had to deal with the much-vexed question of Public Lighting. The greater portion of the information contained in these pages has already been given by him in the form of letters, more or less lengthy, in reply to questions. It may be that many of those letters have either got inadvertently mislaid, or become the victims of the waste paper basket; so that the information, given with some amount of trouble, and received perhaps not without appreciation, has accordingly been lost.

In this pamphlet the Author has endeavoured to arrange in a connected form a quantity of information, which the long study he has given to the question emboldens him to say will be found of great practical utility.

The systems indicated are possibly not everything that may be desired, but they will at least serve as starting points from which to arrive at others which shall be found still nearer perfection.

To this end the energies and skill of the Author in this branch of the science of Gaslighting are devoted, and, although he may not directly succeed, yet that success which is sure ultimately to be obtained is at least accelerated by the contributions he has made towards it.

5

DESCRIPTION OF LAMP GOVERNORS.

(PLATE 1.)

(A) is a steatite burner tip, and the diagram shows the method by which it is fastened into the cone (B), which is screwed to the governor by an ordinary ⅜ gas thread.

(C) is a metal ring which, being screwed into the body of the governor, holds the leather firmly to its seat. Between the leather and the metal ring is a turned card washer, which prevents the leather from being injured by the turning of the screw in tightening it down to its bearing.

(D) is the regulating valve, fixed to the leather (E), by means of two shields (FF), which clip the leather between them, these shields being held firmly by a screwed brass nut (G) on the top of the upper shield, and a tin washer on the under side of the lower one. The leather (E) answers the same purpose in the dry as the gasholder does in a wet governor, rising and falling accordingly as the pressure is strong or weak at the inlet L.

(H) is the junction gasway connecting the outlet of the lower part of the governor to the cover through which the gas passes up the cone (B) to the burner, in the direction shown by the arrows.

(II) are screws which hold the gasway to its seat. Between the gasway and the seat is interposed a washer made of paper, which being painted with a little red lead and oil, secures the soundness of the joint.

(K) is a screw for holding the top securely and preventing damage to the gasway should the lamplighter strike the cone too hard with his torch in lighting, or catch the burner with his sleeve in cleaning the lantern.

(L) is the inlet of the governor screwed to the ordinary $\frac{3}{8}$ gas thread, and the direction of the gas is shown by arrows.

(MM) is an annular leaden weight by which the pressure to be maintained at the point of ignition is fixed. As in the ordinary wet governor a lighter weight will give less pressure, and vice versâ.

(N) is a hole communicating with the external air for the purpose of allowing the leather to rise and fall.

Note.—If this hole becomes stopped, the governor will cease to act. The leather being sound, no escape of gas can take place from here when the governor is in operation.

General Remarks.—To ensure success in lighting

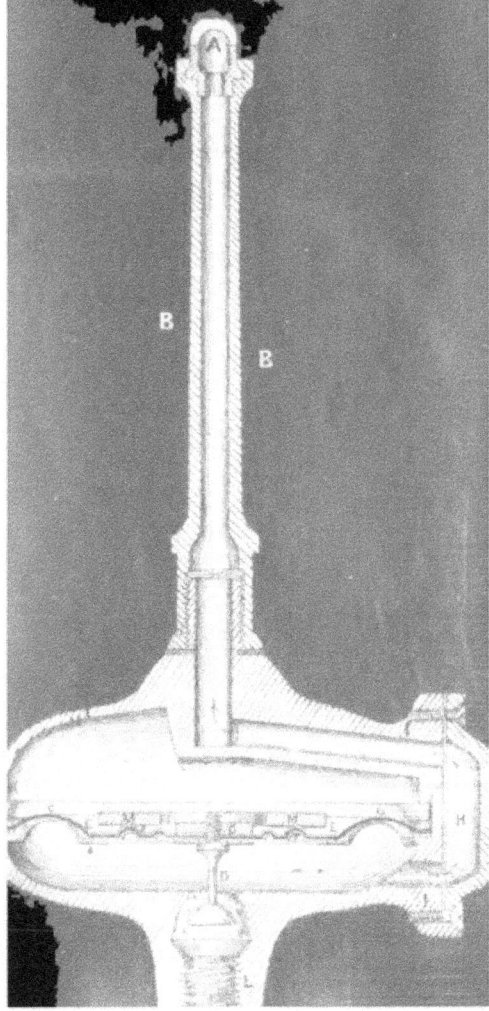

WILLIAM SUGG'S
PATENT
PUBLIC LAMP GOVERNOR

street lamps upon the governor system, the burner
and governor must be considered as one, and any
accident which happens to either, ought to be followed by the removal of both for readjustment, as
described in the following pages.

No watchspring or hard substance should be made
use of to clear the burners, because as there is no
corrosion to remove but simply dust, a brush or piece
of paper will be sufficient for the purpose. The effect
of any attempt to clean with watchspring or hard
tools will probably be either damage to the burner
or its total destruction. It is evident that unless the
material of which the burner is composed is capable
of resisting enlargement, the maintenance of the
regulated consumption at the burner must soon be
at an end, and the inevitable loss of gas would in a
short time be equal in value to the cost of a new
governor and burner.

With careful cleansing, in the manner suggested,
the burners will remain in order for many years.

Note.—Some of the old governors will be found to have a brass
ring instead of a patent metal one for the same purpose. This new
ring, and the method of securing the leather, is the subject of a
second patent, and will be found to possess the merit of greater
simplicity.

APPARATUS FOR ADJUSTING AND TESTING LAMP GOVERNORS.

(PLATE 2.)

THE requisites for this purpose are:—

An experimental meter (A) made in the simplest manner with a measuring drum holding one-twelfth of a foot, and equal per revolution to the rate of 5 cubic feet per hour, into which number the dial is divided, each foot being subdivided into **tenths of a foot.**

A minute clock (B).

A King's pressure gauge (C), capable of showing 4 inches of pressure.

Another (D), **capable of** showing $1\frac{1}{2}$ inches of pressure, **with** subdivisions into hundredths of an inch.

Two double dry governors (E & F), the outlets of both being connected by means of a T piece to the inlet of the meter (A).

A float fitted with 3 lamp cocks (G H & I).

A bent brass tube (K), fitted to take a burner and cone similar to that made use of in the governors, **and** with a connexion tapped **into** the square of the

elbow, so as to communicate by means of an india-
rubber tube the pressure of the gas after passing the
governor to the delicate King's pressure gauge (D),
fixed on the right hand of the meter. (This pressure
is considered to be that which would be found just
inside the orifice of the burner, and is spoken of as
the pressure at the point of ignition.)

A brass **T** piece (L) screwed $\frac{3}{8}$ inch inside at
the bottom and $\frac{3}{8}$ outside at the top, so that it may be
inserted between the cone and case of a governor,
which it is requisite to test in the readiest manner
without disturbing the joints.

FIXING THE APPARATUS.

The best method of fixing this apparatus is that
shown in the drawing annexed. The pressure gauges
(C & D), and the governors (E & F) rest upon brackets
let into the wall, so that their action may not be in-
terfered with by vibration.

The inlets of the 2 double governors (E & F)
should be connected to the outlet-pipe of that gasholder which gives the most uniform pressure on the
works. If the apparatus be not fixed on a gas works,
then it is necessary to have a common holder on
purpose.

Although lamp governors are made for the purpose of maintaining by their action uniformity at the

burner under varying pressures in **the** street mains, yet they themselves cannot **be well** adjusted unless there is a possibility of starting from a standard pressure which can be relied upon to remain unchanged during the whole process of adjustment.

At the **same** time it is necessary, **in** order to prove the delicacy of the governor, to have a means **of increasing the pressure at its inlet** by degrees **till the** highest pressure is reached, which being shut off, the standard pressure resumes its influence.

To maintain this standard pressure is the province **of the** double governor (F), fixed on the right of the meter, **while that on** the left (E), serves to regulate the degree **of** variation in the pressure which **it is intended to cause at** the inlet of the **lamp governor** under examination.

ADJUSTMENT OF THE APPARATUS.

Having arranged the apparatus in the order **pointed out, see that** the pointers **of** the pressure **gauges are at zero** when the gas **is** turned off.

Note.—For this purpose open the blow-off cocks (AA), and shut the gas-cocks (BB).

Adjust **the (F) or** standard pressure governor till **it gives a** pressure upon the **(C)** pressure gauge equal **to the lowest or day** pressure in the mains (say six- **tenths).**

Weight the test or store gasholder (if it is not possible to obtain the pressure from a large gasholder, or in any other way), up to 2 inches or $2\frac{1}{2}$ inches.

For all the purposes of testing, excepting for soundness, a pressure of 2 inches is amply sufficient. To increase the pressure gradually from 6-tenths to $1\frac{1}{2}$ inches is a severer test for a governor than suddenly to change it from 6-tenths to 6 inches.

Adjust the inlet holder of the high pressure governor (E) to 2 inches, or nearly the pressure of the test gasholder.

NOTE.—In adjusting these governors (which are double), the inlet holder must be weighted to give about two or three-tenths more than the outlet holder; thus, supposing the operator were adjusting the low pressure governor (F) to give six-tenths pressure, he would adjust the inlet holder to eight-tenths and the outlet holder to six-tenths. This is done to allow for any variation of pressure which may occur at the outlet of the first holder, and in order that the second may be enabled to maintain at its outlet a perfectly uniform pressure.

The apparatus will now be in order for testing.

Supposing it is required to test a governor taken from, or ready to be fixed in a lamp, screw it on one of the cocks as at (G), commence by turning on the high pressure cock (E E), and light the burner.

NOTE.—This is done to put the governor in action in case it may have lain by for some time, and have become stiffened by reason of the oil from the gas being viscid. If put in action for a short time, the gas will itself restore the leather to its former condition.

Turn off the high pressure cock and turn on the low (F F) without extinguishing the flame, and let it

burn for at least one minute, then, when the hand of
the meter is at five, start the minute clock (which
should be at zero), and at the moment when the bell
strikes one minute, notice the position of the hand
of the meter. The distance travelled over by that
hand in one minute equals the rate per hour at which
the gas is consumed by the burner; thus, if the hand
has made the complete circuit of the dial, the consumption of gas is five cubic feet per hour, if from
zero to 1, one foot per hour, and so on.

If it be necessary to ascertain whether the
governor under examination varies under different
pressures, then unscrew the cone and insert the T
piece as at (H), connecting the outlet of the T to the
point of ignition gauge (D).

Commence with the low or standard pressure,
and ascertain the rate per hour, after which turn
off the low and turn on the high pressure cock
as nearly simultaneously as possible, so as not to
extinguish the light, then gradually increase the
weight upon the governor (E) until the pressure
gauge shows that the highest point to which the
governor is weighted is attained.

Note.—The weight must be put on the outlet holder; after
the first adjustment the inlet holder must not be touched, unless it
is desired to increase or diminish the standard maximum pressure.

By counting the number of hundredths of an inch
of variation shown by the pressure gauge, that of the

governor may be accurately ascertained, and by observing how much the consumption is increased per hundredth of pressure, a table of variations for that particular size of burner may be readily made.

READJUSTMENT OF GOVERNORS.

Supposing a governor consumes more or less than the quantity required, or has become defective in any way, proceed as follows:—

Commence by screwing into the outlet of the governor, in the manner shown in the drawing at (I), the bent brass tube described on page 8, and connect the india-rubber tube with the pressure gauge (D).

Submit it to the variation under the pressure test, as described on page 12. If the amount of variation is within the allowed limits, then proceed to regulate the rate of consumption by increasing or diminishing the weight on the shield (F). *See* drawing of governor, Plate 1.

The burner upon the governor must be lighted and allowed to burn for at least $1\frac{1}{2}$ minutes before testing its rate of consumption.

The adjustment having been completed, replace the cover and carefully make the gasway secure, by making use of a paper washer dipped in grease. Then make a final test for variation and consumption, in order to ascertain that nothing about the cover or gasway interferes with the proper working of the governor.

NOTE.—The paper washers can be supplied ready for use, at 1s. 3d. per gross.

METHOD OF PUTTING IN NEW LEATHERS.

If the governor will not work well, the leather must be examined, and if soft and flexible the defect may be looked for in the valve and seat, which may have become dirty or even partially stopped with naphthaline or other deposit. To remove the valve, a piece of solid brass screwed $\frac{3}{8}$ to fit the inlet of the governor must be carefully screwed up until it just lifts the valve and holds it to its seat without bruising, then the top nut (G) can be unscrewed by the aid of a pair of small pliers. This being removed, the valve will drop out. Then proceed to remove the leather by means of a proper tool for the purpose, which will fit the projection on the metal ring (C). The leather and card washer placed above it will then easily come out, but care must be taken not to pull the leather out of shape more than can be helped. The valve and case may then be boiled in potash water and thoroughly cleansed. After which, proceed to put the governor together again in the following manner :—

NOTE.—A new leather can be put in if necessary. They are supplied in tin boxes ready for use at per doz.

Turn it upside down and drop in the valve, screw in the piece of brass and lift the valve to its seat with just sufficient force to hold it there. See that the rod of the valve is as nearly central as possible with the case, and replace upon it the small tin disc and paper washer which were found upon it when taken to pieces, and upon the latter place the lower shield. Then take the leather and drop it into the case, and with a small piece of wood or bone (a lucifer match, cut to something like a chisel point, answers the purpose very well) dispose it carefully on its seat, taking care that none of the leather comes up alongside the screw. It will be found exactly to fit the case. Put on the upper shield, and screw the top nut (G) with a paper washer dipped in red lead, placed under it, nearly home. This nut must be left loose till the adjustment of the governor is completed. Drop the card washer on to the leather and replace the ring. Very little force need be used, for if laid right it will screw smoothly into its place with but little pressure.

Blow into the outlet and see that the leather plays freely with a stroke of about $\frac{3}{32}$ of an inch.

Having reinstated the leather and valve, place the governor on the test apparatus, and test for variation of pressure as before described. If the variation is

Note.—Care must be used in putting on the shields that the burr, if there is any, must be upwards on the top and downwards on the bottom shields, so as not to hurt the leather.

more than two hundredths of an inch the valve or leather must have been placed in badly, or the leather requires fresh oiling. If it requires oiling time must be allowed for the oil to be absorbed before again testing, say 10 or 12 hours. If the valve has been put in out of the centre it must be put right by shifting till it is so, after which it may be screwed down tightly. When the adjustment is completed the valve rod must be securely fastened in the same manner as it was found, or the vibration in the street will loosen it so that the valve will fall out and probably the governor be destroyed.

NOTE.—A preparation for oiling the leathers when they require it, can be forwarded, packed in tin bottles, price 2s. 6d. per lb. No other material will answer the purpose so well.

METHOD OF FIXING GOVERNORS IN THE PUBLIC LAMPS FOR THE SYSTEM OF TORCH LIGHTING.

(PLATE 3.)

THE following, though not suitable to every locality where it may be required to fix governors to the public lamps, will, however, be found to be most advantageous in the majority of cases. It is based upon the experience gained since the introduction of governors by those who have had them in constant use in Great Britain and Ireland, and various parts of the world.

LANTERNS ON COLUMNS.

These are best fixed in the manner shown in drawing (Plate 3); the cock (B) or (C) being just under the lantern, the governor will cover the hole where the supply enters, and prevent the sharp current of cold air, which always rushes in as soon as the lamp is lighted, from striking the flame, destroying its shape, and otherwise preventing the proper combustion of the gas.

B

The governor being fixed in this manner, the burner will, in an ordinary 14-inch street lantern, just reach the middle of the side pane, which will be found to be the most convenient height.

For lanterns of a larger size, a longer cone must be employed.

Note.—The hole in the bottom of the lantern is not required to be larger than ⅞ of an inch.

In cases where the space between the bottom of the lantern and the top of the column will not be sufficient for the lever cock (B), No. 361 in Catalogue, the improved patent No. 362 (C), which turns on and off by a horizontal movement of one quarter or one eighth of a circle, will be found to work well.

This latter is made of gun metal with a barrel of a hard white amalgum, so as to avoid as much as possible any liability to stick fast. From the fact of the barrel working on the plug in a vertical position, it will readily be seen that there is a certainty of its remaining sound for a longer time than when, as in the ordinary way, the plug itself works in a horizontal position.

The system above given entails frequently the necessity of cutting and rescrewing the stand pipes, which, however, by the aid of proper tools, can be readily done in situ without disturbing the ground.

Upon this point, it may be as well to mention that in all cases the stand pipes ought to be wedged

Pl iii

PUBLIC LAMPS FITTED FOR TORCH LIGHTING

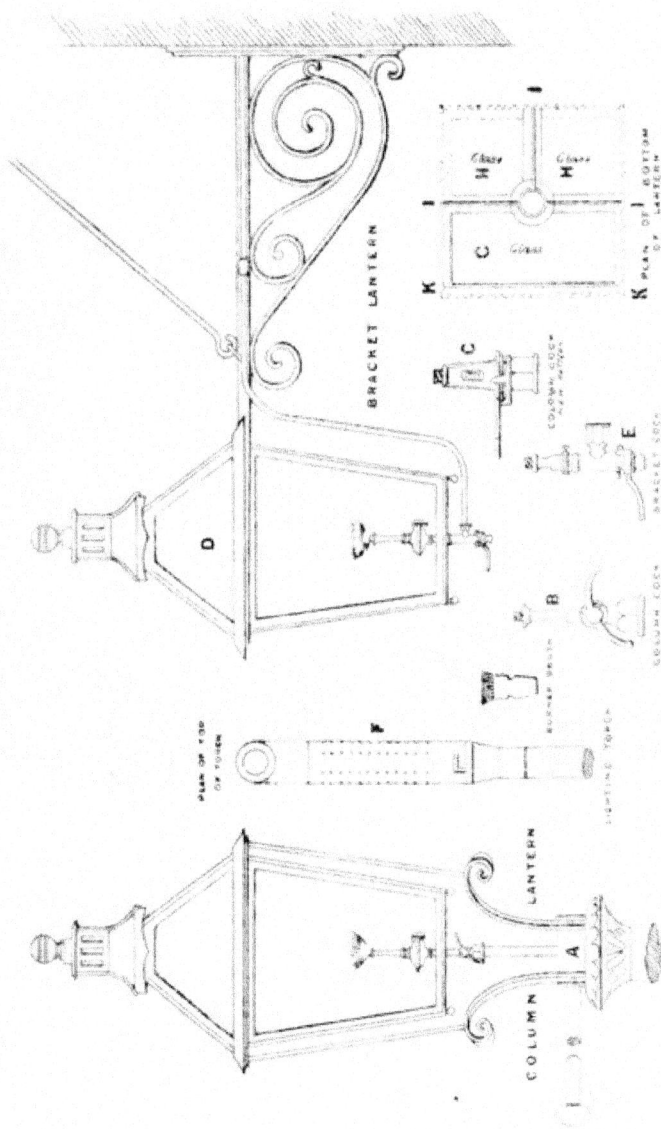

COLUMN LANTERN

A

BRACKET LANTERN

D

PLAN OF TOP OF TORCH

LIGHTING TORCH

F

BURNER GUARD

COLUMN COCK

B

COLUMN COCK FOR LANTERN

C

BRACKET COCK

E

PLAN OF BOTTOM OF LANTERN

K

Glass
H

Glass
H

Glass
C

tightly in the **top of the** columns, for **the continual** vibration they **are subject** to, when **not so treated,** seriously interferes **with the soundness of the joint at** the bottom of the post. The difficulty **of getting at** this joint is **a sufficient reason why attention should** be paid to this **detail.**

BRACKET LANTERNS.

Figure (D) in the drawing shows one of the means by which governors may be advantageously fitted to bracket lanterns. It is a sketch of the most commonly employed style of bracket, and it has been usual to convey the gas into the lanterns by means of copper and sometimes metal tubes or "twiggings," as they are technically termed. The inconvenience and extravagance of this system, from the frequent derangement of the conduits and their almost constantly leaky state, have led to the substitution of iron pipe so fixed that it is not necessary to disconnect the fitting in order to remove the lantern.

The cock (E) made use of is No. 361a in the Catalogue, and is made, the barrel of brass and the plug of gun metal, turning on and off, with a horizontal motion of the lever of one quarter of a circle.

NOTE.—In screwing on the cock, care must be taken that the thread fits tightly, or there will be a possibility of its being turned a little by the frequent stroke of the torch in lighting and extinguishing.

TORCH LIGHTING AND EXTINGUISHING.

This system which, since its introduction into England a comparatively short time since, has been constantly increasing in favour, is however no new thing, for on the Continent it has been practised for a very long time.

However it may have there succeeded, it was nevertheless clearly unsuited to the system of lighting adopted in this country, until the application of governors and steatite burners to street lamps. With these improvements it is found to work uncommonly well, because a simple movement is all that is necessary to turn on the gas, which requires no adjustment, and if the lamplighter brushes his burners when he cleans the lamps, he may confidently depend upon the shape of the flame being always good. The turning off of the gas has for many years been done almost everywhere with the aid of a stick, and to this cause may be traced the almost universal loss of one or more of the quarter bottoms which ought to be found in every street lantern.

A proper amount of ventilation is necessary, or the lanterns would soon fall to pieces from the heat generated by the combustion of what is, considering the size of the chamber in which it is consumed, a very large quantity of gas; but no one who sees the flames blowing about in the manner they do can reasonably say that the best way of insuring that result is attained by the absence of the bottom of the lantern. It is very probable that the loss of gas and consequent loss of light caused by this strong draught would compensate in a very short time for the expense of putting in the bottoms and keeping them in proper repair. In Paris the Gas Company has become so strongly convinced of this fact that they have gone to a very considerable expense in fitting to the lanterns an arrangement of plate glass mounted upon a brass swivel fixed on the supply pipe, which is opened and closed by the lamplighter when he lights the lantern, the hole through which the supply enters the lantern being carefully stopped, if not by the arrangement just spoken of, then by putty.

The torch-lighting system requires but little explanation, and is as follows, viz. :—

The lamplighter is provided with a torch (F), which is simply a small lamp in a brass case, the top part of which is drilled full of holes to admit air to the flame, but so guarded by an inner screen that the wind or a violent motion in carrying it about will not extinguish it, and mounted upon a light staff, varying in length, and jointed or not as may be required. With this he hits the lever of the cock on one side to turn it on, and the other to shut it off. There is a little hole at the top of the torch which projects a ray of light upon the lever sufficient to enable the lighter to find it upon a dark night.

NOTE.—On rainy nights the torch should not be carried perfectly upright, or a drop of rain falling into this small hole may extinguish the light.

Immediately the cock is turned on the torch slips off the lever, and striking the glass flap (G), which is hung on a piece of brass tube soldered along the back of the frame into which it fits, lifts it, and passing up ignites the gas immediately; the torch being withdrawn the flap falls of itself.

NOTE.—These flaps may be made of 20 oz. glass, and will last a very long time. They do not readily break even with a violent blow. They are slipped into grooves and held by a copper tack at each end. The flap bottom complete, as shown at (G, H, H), may be easily applied to existing lanterns by simply cutting out the old cross bottom and soldering the new one at three points (I, I, I) outside the bottom of the lantern; the wire for the flap hinge is put through two holes drilled one on each side of the bottom frame at (K, K).

It is obvious that, to ensure the success of the system, and prevent loss of gas, it is essential that the stops of the cocks should be sufficiently strong to withstand the continual blows of the torch.

ADVANTAGES TO BE DERIVED FROM THE ADOPTION OF STREET LAMP GOVERNORS AND INCORRODIBLE BURNERS.

In order fully to appreciate these advantages it is necessary to commence with this fact, that every 5 cubic feet of gas consumed ought to give the full illuminating power of 12, 14, or 16 sperm candles, burning at the rate of 120 grains per hour, as it does when consumed in the testing burner made use of to verify the illuminating power of the gas made.

Experiments made with burners taken from public lamps in various places usually show that when regulated to consume exactly 5 cubic feet per hour, the amount of light is not quite half that to be obtained from the same quantity of gas consumed in the testing burner. This circumstance has always been a prolific source of trouble to manufacturers of gas, because the commercial public naturally consider that, if they pay for 5 feet per hour of 12, 14 or 16-candle gas, as tested and declared to be such by the appointed gas testers, they ought to see as much light given from it as could be obtained from the corresponding number of sperm candles. Satisfied that they do not get this amount of light they immediately jump to the conclusion that the gas company or manufacturer does not supply the lamps with 5 cubic feet per hour, and up to this time no amount of argument has sufficed to convince them of their error, for error it undoubtedly is, as any one, who carefully examines the question, must see. Thus, to consume an average rate of 5 cubic feet an hour in a public lamp, subject to the ordinary variations of pressure consequent upon the constantly changing demand upon the street mains, a burner must be selected having apertures so fine that the quantity of 5 feet is blown out at such a rapid rate as to give it no chance of obtaining one quarter of its illuminative power; such a burner, of course, cannot be employed, and for it is substituted one with larger apertures, over which, of course, the variations of pressure have a greater influence. Add to this the constant enlargement of the apertures for the escape of gas by the liberal use of the watch-spring or the broach, and it will be readily believed that in every case where the quantity of gas consumed by public lamps has been impartially and correctly tested, it has been found that the real average consumption has been at least 25 per cent. above that paid for, and yet dissatisfaction has been the constant expression on the part of the consumer.

To this unsatisfactory state of things the adoption

of street lamp governors brings certain and easy
remedy, for, when once the control of the pressure at
which the gas is to be consumed is obtained, nothing
prevents the employment of the most suitable burner
for the proper combustion of the gas, so as to insure the
full illuminative power which it is capable of evolving;
and this burner being made of a material which resists
damp and heat, and does not admit of enlargement by watch-spring or broach, the contract
quantity may be maintained to the satisfaction of
both buyer and seller.

This is abundantly borne out by the experience
gained in a large number of cities and towns in
various parts of the world where this system has
been in operation.

It has frequently been asked, how can it be
guaranteed upon this system that the quantity of gas
contracted for is given during all the hours of the
night? The answer is, very simply. If the pressure
in the mains has been kept a trifle higher than that
required to work the governors, no doubt can be
entertained that the supply continues the same.

Assurance of this adequacy of pressure is obtained
by fixing, in several parts of the town, lighted upon
this system, some good pressure registers. Their
indications will afford a sure basis upon which to
settle terms, in the event of inadequacy and consequent non-fulfilment of the contract.

This completes the system of the supply of public lights with governors alone.

By permission of the Committee of the British Association of Gas Managers, a paper, read by Mr. Thos. H. Methven at their meeting in June, 1868, is here annexed :—

On the necessity and advantages of Regulation of the Gas consumed in Public Lamps.

For many years public lamps have been a much neglected though an important part of the business of gas companies. The object of this paper is rather to draw attention to the subject than to introduce anything new in connexion with it.

Several reasons may be given for the neglect and apathy that have so frequently prevailed. The chief one is, that the contract system so frequently allowed so small a profit that there was no inducement to improvements that entailed any cost. The lamps were therefore lighted and extinguished at the appointed hours, and few inquiries were made by gas companies regarding the quantity of gas consumed in them, neither was much care bestowed upon the best method of getting the full advantage of the light obtainable from the gas.

The earlier lamps were badly constructed and small in size, yet the tinwork was frequently so wide as to obstruct the light which would otherwise have passed through the small and dirty panes of glass. To have removed the whole and have substituted larger and better made lamps would have been expensive, so the old pattern was adhered to, even when additions to the number were made.

The lamplighter was instructed to adjust the flames to consume 5 feet per hour. However anxious to obey his instructions, he seldom could do so. Sometimes, in order to economise gas, burners with thin slits were used, which only allowed a consumption of 5 feet per hour at the maximum pressure; of course, much less was consumed towards midnight. This burner wasted a large percentage of light, and gave rise to repeated complaints.

With ordinary burners it was quite true that all the lamps did not consume 5 feet per hour, but many more, and especially those before the lighting commissioners' doors, consumed nearly double the amount. Frequent disputes arose with the public regarding the quantity of gas consumed, in consequence of which occasionally lamps were supplied through meters. These were frequently unskilfully attached, and gave very irregular and unsatisfactory registration on that account. Again, when lighted by the company, the lamps consumed excessively, and when by the lighting commissioners the result was quite the reverse.

All confidence was lost in the lamplighter, and

hence the double tap was introduced. At first a tap was placed in the upright pipe, and adjusted and placed out of the lamplighter's power. Afterwards both taps were made in one casting, one being adjusted by a key so as to allow 5 feet per hour to pass; the other tap was for the lamplighter's use.

This arrangement was a step in the right direction, although the flames were adjusted rudely by a gauge. It should have been introduced before, when a constant pressure was maintained in the mains. I recommend it now to all badly managed gas-works as better than no regulation at all. There are not a few who might at once adopt it. But wherever there is variation of the pressure, say from 20/10ths in the evening to 6/10ths after midnight, there it is useless, for if the double tap allows 5 feet to pass at the one period, it will not allow half that quantity to pass at the other.

The attempts to regulate the consumption by fixed orifices having failed, regulators were introduced. Mercurial ones were the first in the field; though correct in principle, yet in practice they were found unsuitable, being too delicate for their exposed position. Then dry governors were introduced, and to these the writer would draw attention; they have a leather diaphragm, having a conical valve attached to the under side of it, and weighted on the top side to allow the consumption required.

The regulator is fixed inside, or under the lamp, and is furnished with a burner which, at 15/10ths pressure, consumes 9½ feet per hour. It is well known that a wide slit is the best for obtaining the full illuminating power from the gas consumed. At any pressure under 25/10ths the regulators maintain a uniform flame. In experiment I have found them work with 55/10ths, but I am informed by an engineer who has a hilly district—where the pressure is 34/10ths—to supply, that the pressure forces up the valve so close to the seat that an insufficient light is obtained. This defect might easily be remedied by fixing before the ordinary regulator another adjusted to give 20/10ths pressure.

Not only is there a great advantage to the gas companies resulting from the use of the regulator, but there is also a better light supplied to the public. This arises from the superior description of burner that can be used, which is usually supplied by the manufacturers of these instruments. The lamps, however, require to be more carefully constructed than formerly, as, on account of the low pressure at which the gas is consumed, there is a greater tendency to unsteadiness and smoke in the flames on windy nights.

The great difficulty in adopting improvements is in the large cost involved; yet, if a plan were resolved on, it might be carried out piecemeal at first, when lamps were repaired or new ones were erected. Then, by an **extra effort, the whole might** be altered **to** the advantage **of all concerned.**

A bad regulator, **or an inferior one, is not worth** fixing; **but** several makers produce excellent **ones.** A **good** regulator should be easily taken to pieces, so that **new** diaphragms may be inserted when necessary; **it should** have burners **of** incorrodible material secured to it by the maker, **and** it should be tested in this complete state before leaving his **factory.**

That there **is** plenty of room for economy in **the** public lighting **may** easily be **proved** by fixing a few meters **to the lamps in** different parts **of** the town. The meters should be carefully fixed in air-tight boxes. The writer has found the consumption **vary** from 5 to 10 feet, **and in a** somewhat lengthened trial has found **the average, even** with the greatest care in lighting. **6½ feet** per hour.

With a regulator a better **light can be had from** 5 feet. Thus, in a **town** with 300 lamps, lighted **1200 hours** annually. 540,000 feet **of gas** might **be saved.** In addition **to** the pecuniary result, which would **soon** pay **for** the alteration, there would be this advantage, that **the** lighting committees and **the** public would **have** no such grounds of complaint as **too** frequently **at** present exist, **and** much of that ill-feeling and consequent agitation against gas companies which **is so** detrimental to the interests of all concerned would be prevented.

THE SUPPLY OF PUBLIC LAMPS ON THE SYSTEM OF AVERAGE METER INDICATION.

This question has been so fully and ably treated in a paper read before the British Association of Gas Managers, at their meeting in June, 1868, by Mr. Charles Hawksley, that to reprint it in this work will be to give the best description of the system as originated and carried out by Mr. Thos. Hawksley, the past president of the Association, at Nottingham. This being the only instance, up to that time, in which the average meter system has been put in operation with equal advantage to the gas company and the public authorities, the kind permission of the President and Committee has been asked and obtained for the reprinting in *extenso* of the paper.

On the Lighting of Public Lamps on the System of Average Meter Indication.

The necessity for the regulation of public lights having been already discussed in the paper submitted to the Association by Mr. Methven, the present paper will be confined principally to a description of the method adopted in the town of Nottingham for effecting the supply of gas to the public lamps on the system of average meter indication.

Average meter indication appears to have been in use in one or two of the smaller towns in this country for a period of at least sixteen years, but was first brought prominently into public notice when applied in conjunction with the "double tap" in the town of Reading during the year 1863, under the direction of Mr. Samuel Hughes.

The attention of the Nottingham Gas Company was, in the year 1860, called to the excessive quantity of gas consumed by the public lamps, and on investigation it was found that the mean average consumption by each lamp amounted, during the six months ended September 1, 1860, to no less than 7·3 cubic feet per hour, although the contract with the town authorities provided only for 5 cubic feet per hour per lamp; and during a portion of the period referred to—viz., from the 1st to the 22nd of June—the consumption per lamp was found to have attained the enormous quantity of 9·5 cubic feet per hour. The amount of gas actually consumed (as ascertained by meters attached to several of the lamps), when divided by the number of hours during which the lamps ought to have been lighted according to the lighting table, gave the results mentioned above, showing great negligence on the part of the lamp-lighters, who, especially during the height of summer, lighted the lamps earlier and extinguished them much later than the hours stated in the table.

It was to amend the then unsatisfactory state of the arrangements with the lighting authorities of the town that the company applied to Parliament, in the session of 1864, for an Act having amongst other objects the making of better and more effectual provisions with regard to the lighting of the public lamps. The clauses relating to this subject in the bill as introduced into Parliament were as follow:—

16. Subject to the provisions of this Act the Company shall, at their own expense, upon the request in writing of any lighting authority, provide, lay down, fix, maintain, and keep in repair all mains and service-pipes, lamp-posts, brackets, lamps, burners, stop-cocks, regulators, and other apparatus connected therewith, necessary for the proper lighting of such of the streets within the said limits as are mentioned in such request, and provided the lamps to be supplied shall be fixed at not exceeding the average distance of eighty yards along the course of any main to be laid down by the Company for conveying gas to such lamps.

17. The Company shall, at the request in writing of any such lighting authority, supply all or any of the present public lamps within the said limits, or such other public lamps to be hereafter provided and fixed as aforesaid, with so much gas, and to be delivered at and for such times and periods as the parties on whose request the supply of gas is made may from time to time desire.

18. The price to be charged by the Company, and to be

c

34

paid to them by the lighting authority, for all gas so supplied to or for any such public lamps, shall always be calculated and fixed at and according to the lowest sum per one thousand cubic feet for the time being charged by the Company to any **private** consumer in the parish or place in which such public **lamps** shall be situated; but this shall **not** in any manner **alter** or affect any contract for the supply **of** gas to such **public** lamps entered into before the passing of this Act, and **then** subsisting, **or** the price **to be** charged and paid for the gas supplied or **to be** supplied under any such contract.

19. The Company shall light, clean, and extinguish all public lamps to which they shall supply gas, and the sum to be paid to the Company by the lighting authority for such lighting, cleaning, and extinguishing, and for the use, maintenance, and repair of the **service** pipes, lamp-posts, brackets, lamps, burners, **stop-cocks**, regulators, and other apparatus as aforesaid, shall **be sixteen** shillings per lamp per annum, payable quarterly.

20. Notwithstanding anything hereinbefore contained, the **quantity** of the gas supplied to the public lamps under any **contract** hereafter made shall, **at** the option either of the **Company or** of the lighting **authority**, be ascertained during **the period of any such contract** by meter, and in such case the **necessary meters** shall be provided and fixed and kept in repair by **the Company,** at the expense of the party requiring the same.

21. The Company shall, within six **months** after the passing of this Act, pay to the lighting authority such sum of money, by way of **compensation, as may** be agreed upon **between** them as the reasonable va'ue of the lamp-posts, brackets, lamps, **and** the fittings **and** apparatus connected therewith, belonging to and used by such lighting authority; or in case any difference shall arise as to such value, the same shall be settled **by** arbitration, in manner provided by the clauses of "The Companies Clauses Consolidation Act, 1845," with respect to the settlement of disputes by arbitration; and

at the termination of the current quarter in which the payment is made, the lamp-posts, brackets, lamps, and the
fittings and apparatus in respect of which the payment is so
made, shall belong to the Company; and any sum of money
agreed on or awarded to be paid, together with the cost of
the reference (if any) as settled by the arbitrators under the
provisions of the said Act, shall be paid out of the reserved
fund of the Company.

The bill was opposed by the corporation and the
lighting authorities of the town, and after a severe
contest before a select committee of the House of
Commons, during which certain alterations were
made in the bill affecting the regulation of the public
lamps, the Act was passed with the following
clauses:—

15. Subject to the provisions of this Act the Company
shall, at their own expense, upon the request in writing of any
lighting authority, provide, lay down, fix, maintain, and keep in
repair all mains necessary for the proper lighting of such of the
streets within the said limits as are mentioned in such request,
and provided the lamps to be supplied shall be fixed at not
exceeding the average distance of eighty yards along the
course of any main to be laid down by the Company for
conveying gas to such lamps.

16. The Company shall, from time to time, at the request
in writing of any such lighting authority, supply all or any
of the present public lamps within the said limits, or such
other public lamps to be hereafter provided and fixed as
aforesaid, with so much gas, and to be delivered at and for
such times and periods as the parties on whose request the
supply of gas is made may from time to time desire.

17. The price to be charged by the Company, and to be
paid to them by the lighting authority, within the extended

limits by this Act authorized, for all gas so supplied to or for
any such public lamps, shall always be calculated and fixed
at and according to the lowest price for the time being charged
by the Company to any private consumer in the parish or
place within such extended limits in which such public lamps
shall be situated.

18. The gas supplied to the public lamps within the
limits of this Act and the recited Acts shall be consumed by
meter, at the option, from time to time, of the lighting
authority or the Company; and in case of its being consumed
by meter, the meters shall be provided by the Company at
the expense of the lighting authority, but neither party shall,
except as hereinafter provided, be entitled to require that a
meter be affixed to more than one in every twelve lamps then
supplied with gas under this Act or the recited Acts: provided also, that the Company shall be at liberty, if they
think fit, to have a meter affixed to any additional number of
lamps, they providing such meters, and paying to the
lighting authority the additional expense of providing and
adjusting lamps, lamp-posts, and other things necessary for
their reception and use : provided always, that if the gas
shall, under the provisions of this Act, be supplied to the
public lamps by average meter indication, the Company shall,
for securing uniformity of consumption between the metered
and the unmetered lamps, from time to time provide the
public lamps under the control of the lighting authority with
proper regulating apparatus and burners to the satisfaction of
the lighting authority, or, in case of difference, as from time
to time shall be settled by the Justices in Petty Sessions
assembled.

19. The average amount of the indications of all the
meters attached to the public lamps under the control of any
lighting authority shall be deemed to be the amount consumed
by each such lamp.

20. The gas supplied to any such public lamp shall be
permitted to pass unrestricted to and from such regulating

apparatus for the whole of the period during which any such public lamp shall be lighted.

Subsequently, Mr. Hawksley, the engineer to the Gas Company, was instructed to take the necessary steps for the introduction into the town of Nottingham of the system of average meter indication, and with the assistance of Mr. William Sugg, he devised the modified form of meter and the burner-cock now in use, with the view to overcome some of the difficulties which had previously been encountered where that system had been tried. The lamps at Nottingham were first lighted on the system of average meter indication in the beginning of the year 1866, since which, and up to the present time, that method of lighting has been maintained in operation without interruption. The apparatus employed is as follows :

1. Every lamp throughout the town is furnished with a brass cock, above which are fixed a governor and steatite burner.

2. One lamp in twelve has, in addition to the above, a wet meter placed underground, near the foot of the lamp-column.

The meter is of the compensating class, and, in order to reduce the friction to a minimum, the drum is made of the same diameter (about 12 inches) as that of an ordinary 5-light meter, but is so diminished in width as to have the capacity of a 3-light meter only; it makes eight revolutions for each cubic foot of gas measured, and requires to work it a pressure of only half a tenth of an inch when passing 5 cubic feet of gas per hour.

The waste-water box (A) is so arranged as to be capable of being emptied by means of an exhausting syringe introduced through a plug-hole (B) in the top of the meter-case.

The index (C) is placed horizontally on the top of the meter, so as to be visible on raising the cover (D) of the cast-iron box in which the meter is placed. The index at first employed was made of brass, and of the ordinary pattern; but great trouble being given by the meters ceasing to register, it was discovered that the condensation in the dial box due to changes of temperature, and probably in some measure also to the evaporation from the water in the meter, corroded the wheelwork to so great an extent as to cause the breakage of the teeth of the wheels, and consequently to permit the passage of the gas without registration. It was then determined to make the indices with strong wheelwork of gun-metal, afterwards tinned to preserve it from corrosion, and a simplified arrangement of index suggested by Mr. Henry T. Humphreys was used, consisting of two large wheels, each about 4¼ inches in diameter; both wheels are worked by the same pinion fixed on a vertical shaft, which is driven in the usual way by a worm on the drum-shaft. One of the large wheels is

THE LIGHTING OF PUBLIC LAMPS ON THE SYSTEM OF AVERAGE METER INDICATION.

PUBLIC LAMP METER AND CAST IRON METER BOX

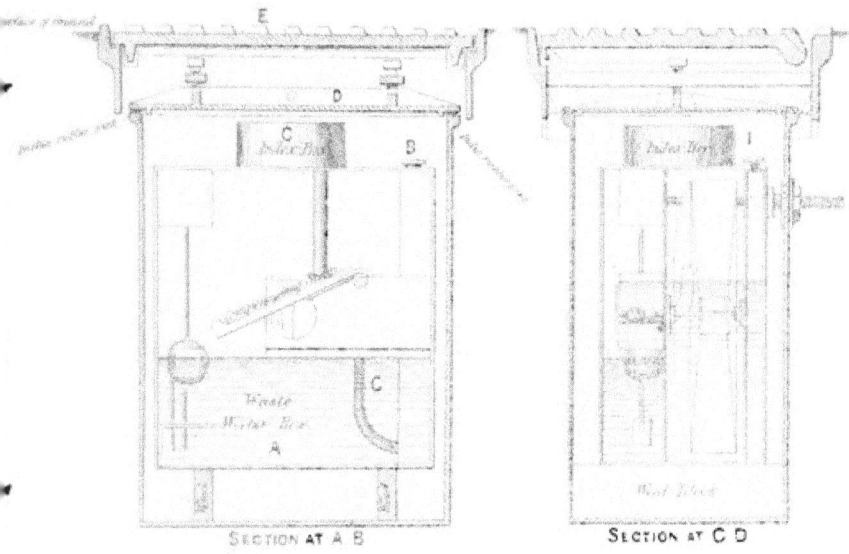

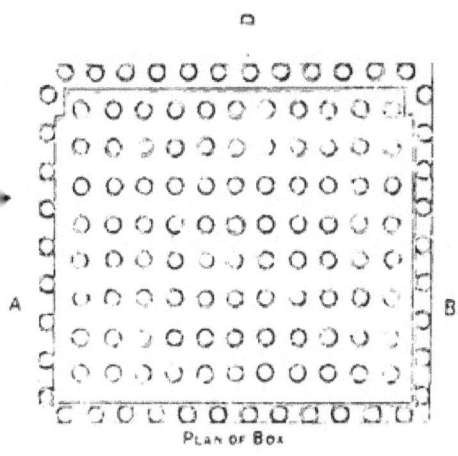

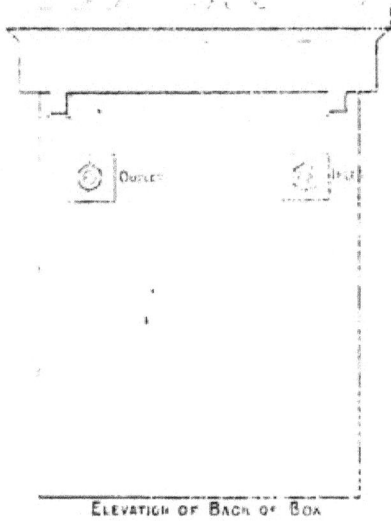

SECTION AT A B

SECTION AT C D

PLAN OF BOX

ELEVATION OF BACK OF BOX

Scale 14 In. 1 Foot

provided with 202 teeth, and is attached to a revolving dial-plate, the circumference of which has 100 divisions, each representing one cubic foot of gas, indicated by means of a fixed pointer. The other large wheel is furnished with 200 teeth, and has attached to it a hand also pointing to the before-mentioned divisions on the dial-plate, each of which now represents 100 cubic feet; but the number of teeth in this wheel being two less than those on the wheel to which the dial-plate is connected, the hand revolves somewhat faster than the dial-plate, and thus indicates every 100 cubic feet of gas consumed up to 10,000 cubic feet, beyond which it is unnecessary to record in meters of this class. As a further precaution against corrosion, the index-box is filled with refined oil to the level of the under side of the dial-plate, and with the index thus made and protected no further difficulty has been caused through the meters having ceased to register.

The freezing of the water during the winter has been effectually prevented by the introduction into each meter of a small quantity of methylated spirit. This was at first found to interfere with the proper action of the meter, and it was then discovered that the spirit as ordinarily sold is mixed by the Customs authorities with gum to prevent its use for the purpose of defrauding the Excise. Unadulterated spirit was afterwards obtained, on a certificate being

40

given to the authorities as to the use to which it **was**
to be applied, and no further inconvenience has been
experienced. Additional protection against frost is
afforded by filling with felt **the space** between the
two covers of the meter-box.

The meter is placed in a cast-iron box let into the
ground near the foot of the lamp-post, so that the cast-
iron lid (E), **which is** hinged, is level with and forms
part of the foot pavement; beneath this lid is a false
cover of iron (D) **to afford** additional protection to the
meter, which cover may, if required, be tightly
screwed down on to an india-rubber washer, thus
forming a water-tight joint **and** keeping the interior
of the box dry even when immersed in water.
Apertures are left in the back of the box, as shown
in section at C, D, through which to pass the inlet
and outlet pipes, and these are made water-tight by
means of washers and **back nuts.**

After having passed through **the** meter, the gas
ascends by a pipe placed in the usual manner in the
lamp-column until it reaches a brass cock (B, Plate 3)
placed between the top of the post and the under side
of the lantern. This cock is opened and closed by
means of two short brass arms, curved downwards, so

as to be readily caught by the end of the lighting torch. The plug is made longer than usual, for the purpose of securing tightness and resisting the blow of the lighting torch when used with rapidity, and the stops are so placed that the plug cannot make more than a quarter of a revolution. This arrangement ensures the opening of the cock to the full extent whenever the lamp is lighted, without requiring care on the part of the lamplighter, who has merely to push up the lever as far as it will travel, and who would, indeed, have some difficulty in opening the cock only partially, were he even disposed so to do.

Immediately above the cock, but inside the lantern, is placed the governor, an instrument without which the system of average meter indication could never have been satisfactorily adopted. The governor in use at Nottingham is the one made by Mr. Sugg, which is so generally known that it is almost unnecessary to state that it is of the kind called a dry governor, and consists of a leather diaphragm from which is suspended a cone which regulates the size of the aperture admitting the gas to the under side of the diaphragm, and which can by means of weights be so adjusted as to give any required pressure at the burner, the stem of which is screwed on to the top of the governor. The burner is a bat's-wing, formed of steatite, and placed at the upper end of a brass stem made of unusual length, with a **view** to prevent the heat of the flame injuring the leather of the governor,* to diminish to an almost imperceptible amount the shadow which would otherwise be thrown on the pavement by the governor, and to obtain a regular flow of the gas before it reaches the point of ignition, a matter of some moment where it is desired to **have a steady flame.**

Meters **are, as has** been already stated, attached **in the case** of Nottingham **to one** lamp in twelve, the lighting authorities **having** elected to have that proportion, although the gas company were willing to **adopt one in** twenty; **it is,** consequently, of the **utmost** importance **that** the governors and burners **of** any one **series of** twelve lamps should each **be** accurately adjusted to consume an equal quantity of gas, otherwise **the** metered lamp would cease to afford **a** correct indication **of the** consumption of the whole of **the** remaining eleven lamps. The governors, **with** their burners attached, are therefore, in the first instance, separately adjusted **to a** consumption of 5 cubic **feet** of gas per hour; they are then placed twelve in a row, and are again tested for an hour, when, if **the** total consumption during that time is found **to** be 60 cubic feet, they are issued for fixing

* It is found in practice that even with a very short cone the heat does not affect the leathers injuriously.—W. S.

to the lamps, the governor and burner for the metered lamp being taken indiscriminately from the set of twelve. Care is, however, taken to keep each set distinct, and in the event of any future readjustment or repair being required to the governor or burner of any one of the set, the remaining eleven are also removed from the lamps, and the whole are again tested together before being refixed. With these precautions no difficulty is experienced in maintaining uniformity of consumption in both the metered and the unmetered lamps.

Having described the apparatus in use at Nottingham, it will now be shown with what results the application in that town of the system of average meter indication has been attended. The public lamps have been supplied on that system for nearly $2\frac{1}{2}$ years, but as some months of that period were necessarily occupied in bringing the system into perfect operation, the results given will be confined to the eighteen months ended March 31, 1868, that being the date of the last quarterly return of the indication of the meters.

There are within the town of Nottingham about 860 lamps, to 72 of which meters are attached; this is exclusive of places without the town in which the system is in operation, and where the results are similar to those obtained within the town. The indices of the meters are recorded monthly, and

returns are made both monthly and quarterly showing the situation of each metered lamp, the state of
the index, the number of cubic feet of gas consumed
during the period, the number of hours during which
the lamp was lighted, and the consumption per hour
by each lamp.

The following table shows the results of the
eighteen months' working. In order to economise
space the indications of seven only out of the 72
metered lamps are tabulated in detail; it should,
however, be stated that the instances given have not
been specially selected, but that every tenth metered
lamp has been taken for illustration exactly in the
order in which it happened to stand in the quarterly
return of the company, as would be at once apparent
by a comparison of the table with the company's
printed form of returns:—

TABLE *showing the Results of the System of Average Meter Indication as applied to the Public Lamps within the Town of Nottingham, during the Eighteen Months ended March 31, 1868.*

Number of Metered Lamps	Situation	Consumption in Cubic Feet per Hour for the Quarter Year ended						Consumption in cub. ft. per Hour at each Lamp on Average of 18 Months.
		1866.	1867.				1868.	
		Dec. 31.	Mar. 30.	June 29.	Sept. 30.	Dec. 31.	Mar. 31.	
10	Newdegate Street	4·54	4·74	4·86	4·71	4·56	4·72	**4·68**
20	Park Row	4·15	4·37	4·49	4·55	4·47	5·00	**4·50**
30	Wilford Road	4·69	5·02	5·05	5·20	4·88	4·90	**4·95**
40	Canal Street	3·70	4·19	4·49	4·87	4·56	4·45	**4·37**
50	Robin Hood Street	4·23	4·46	4·68	5·20	4·64	4·63	**4·64**
60	Vicarage Street	4·23	4·56	4·86	4·87	4·68	4·63	**4·67**
70	Stoney Street	4·38	4·56	4·68	4·55	4·31	4·27	**4·45**
	Average consumption per hour of the above 7 lamps during each quarter year	4·27	4·55	4·73	4·85	4·61	4·65	**4·60**
	Average consumption per hour of the 72 metered lamps within the town of Nottingham during each quarter year	4·32	4·64	4·71	4·86	4·69	4·68	**4·65**
	Maximum consumption per hour of any of the 72 metered lamps within the town of Nottingham during each quarter year	4·85	5·3	5·24	5·36	5·20	5·27	—
	Minimum consumption per hour of any of the 72 metered lamps within the town of Nottingham during each quarter year	3·70	4·09	4·11	4·22	4·15	4·27	—

It will not fail to strike the members of the association how closely the average results of the 18 months' consumption of gas by the seven lamps given in the table in detail corresponds with the average consumption during the same period of the whole of the 72 metered lamps within the town, being in the one case 4·60 cubic feet per hour, and in the other case 4·65 cubic feet per hour, showing that almost precisely the same result would have been arrived at had the meters been applied to 1 lamp in 120 instead of to 1 in 12. Indeed, had it not been for some unusual irregularity in the burning of metered lamp No. 40 during the quarter year ended December 31, 1866, the average hourly consumption of the seven lamps would probably have been exactly equal to that of the 72 lamps. It would, however, in practice probably not be prudent to attach meters to fewer than 1 lamp in 25, in order not only to obtain security against fraud on the part of the lamplighters, but also to give satisfaction to all parties.

Since the preceding paper was read, the system detailed therein has been in operation up to the present time, and is still so continuing. The results have been of the most satisfactory kind.

By the aid of information derived from copies of returns supplied to the Nottingham Gas Light Company, and kindly lent to the author by Messrs. T. & C. Hawksley, a diagram has been constructed which shows with what almost clockwork regularity the system has been worked during a period of 5 years since it was first put in operation.

The diagram will be found at page 67.

FIXING THE PUBLIC LAMP METER.

The cast-iron box which is to contain the meter should be sunk into the ground at the foot of the column on that side facing the footway, the cover forming part and being level with it. It is roughed upon the top, so as to render it safe for passengers.

It is generally considered by those who have had much experience of this system, that the service-pipe from the main to the meter box is best made of lead,—a wooden trough being laid under the tube to keep it from sinking. There is no harm done if the fall of pipe is towards the meter.

The outlet-pipe should be in lead, and the stand-pipe in iron, as usual.

The stand-pipe should be supported in such a manner that its weight is not upon the outlet of the meter. This is easily done by making the pipe longer than usual, and allowing the end of it, which should be capped, to rest on a stone placed inside the base of the column and firmly rammed into the earth.

A T-piece inserted at the proper height receives the outlet pipe from the meter.

The meter, when fixed in the box, should be levelled.

The levelling points are across the meter in front, and on the right side of the index-box.

If levelled up in this manner, it will stand exactly in the position in which it was tested.

METHOD OF FILLING AND ADJUSTING THE WATER-LINE IN PUBLIC LAMP-METERS.

AFTER removing the lid of the box, take out the plug (I, Plate 4). Pour in the water from here till it is presumed that the meter is quite full,— *i.e.*, the measuring-chamber filled to its proper water-line, and sufficient allowed to overflow so as properly to charge the waste-water box.

Then insert the long tube of a syringe similar to this,—

at B, and draw out as much water as possible.

It will readily be perceived that as soon as the tube (G) is unsealed, the syringe will cease to draw water, and the waste-water box will be left full.

Sufficient water to seal the tube can then be added, and the meter is then properly charged. Replace the plugs and it is ready for work.

It is not necessary to have the tube (G) in the

D

meter at all; some have merely a hole in the bottom of the side tube, at the point where the present water-line tube (G) enters it. In this case, the long tube of the syringe is pierced in the side, at a point corresponding with the height of the water-line when the waste-water box is full.

It is obvious that by this arrangement sufficient water must always be left when the syringe ceases to draw.

In the front of the meter will be found a plug at the full water-line of the waste-water box. This is merely for the use of the official tester, and cannot be got at when the meter is at work in the street.

The capacity of the measuring drum of these meters is 1-12th of a cubic foot per revolution.

The average friction of the meter in work, passing gas at the rate of 5 cubic feet per hour, is equal to the pressure given by a column of water half a tenth of an inch in height.

The index-box may be filled with best almond oil; doubtless any oil will do, such as paraffin or petroleum, which will not thicken, dry up, or freeze hard.

When all this is done, the top must be well screwed down to prevent water from entering.

METHOD OF READING THE INDEX.

The index (*see* Plate 5) is of the kind called differential. The index-hand and dial both revolve; but, as a consequence of the hand being fixed to a wheel having 200 teeth and the dial to another having 202 teeth, both of which are driven by the same pinion, the hand gains on the dial two teeth every complete revolution of the latter, which occurs when 100 cubic feet of gas have passed through the meter. Thus it follows that each of the smallest divisions is equal to 100 feet, by the indication of the centre hand (A), and one foot by that of the fixed pointer (B).

When the former has made a complete revolution of the dial, 10,000 cubic feet of gas will have been passed through the meter.

The reading of the index shown in the diagram is 8,575½ cubic feet, because the pointer stands at between 85 and 86 and between 8 and 9 thousand. The hundreds being shown by the small divisions,— of which five are complete, the sixth being incomplete,—are read from the pointer (B), which is at 75½ cubic feet.

It will readily be seen that this is the easiest index possible to read, because the entire quantity consumed may be read with sufficient accuracy by noting the position of the centre hand (A) only.

Care must be taken in the event of this being between two numbers that the lowest be taken as thousands.

The small dial showing 5 cubic feet per revolution is only of use to the official tester. It is fixed on the pinion and revolves with it.

THE LIGHTING OF PUBLIC LAMPS ON THE SYSTEM OF AVERAGE METER INDICATION

INDEX FOR PUBLIC LAMP METER

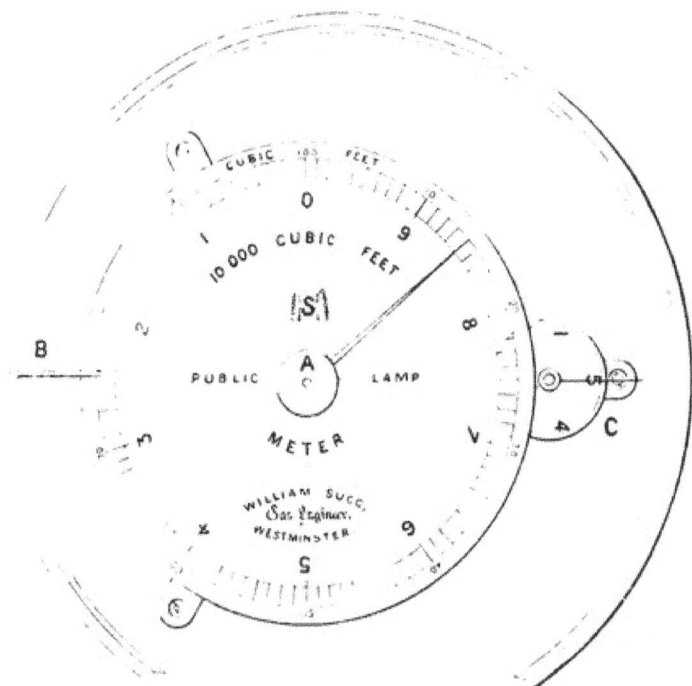

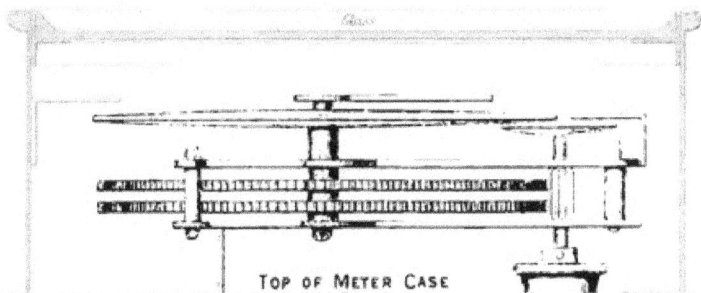

CUBIC 100 FEET

0

10 000 CUBIC FEET

[S]

PUBLIC A LAMP

METER

WILLIAM SUGG,
Gas Engineer,
WESTMINSTER

B

C

Glass

TOP OF METER CASE

Scale ½ Full Size

GENERAL REMARKS.

In order to ensure the success of this system, great attention must be paid to the soundness of the pipes and connexions of the metered lamps, especially that leading from the meter to the lantern. Every leak in the pipe, cock, or governor is multiplied as many times as there are lamps represented by the metered lamps, *i. e.*, if there are 12 unmetered to one metered lamp the error is multiplied by 12, and so on.

Another important point is that the lamps should be lighted and extinguished at the proper time, care being taken that the cocks are effectually opened and closed.

The arrangement of the stop affords a ready means of verifying this fact to the lamplighter.

The recommendations to be found under the head of "General Remarks," page 56, must be attended to.

Thus the much-vexed question of Public Lighting may be satisfactorily arranged and troublesome disputes avoided, at a cost considerably less than that incurred very often in one year by litigation.

In every case, the general determination of the seller of gas to act up to his contract is effectually shown, and distrust on the part of the buyer is entirely removed.

LANTERNS.

HAVING now arrived at the conclusion of the remarks upon the regulation and measurement of the supply of gas to public lights, it must be observed that there still remains an element of discomfort which cannot be obviated without the adoption of a system, instead of that want of it which at present prevails. Much thought has already been bestowed upon the subject of suitable lanterns, but as its consideration involves so many points of taste and design, it will be best in this work simply to define the principles which admit of most easy application to differing circumstances, rather than to discuss the merits of the many kinds of lanterns now before the public.

The requisites for a good serviceable street lantern may be summed up as follows, viz:—

Strength, and at the same time lightness, throughout the framework. As much glass as possible, firmly secured. Ventilation sufficient, and so arranged as to avoid down draught as well as too rapid a rush of air from the hole round about the stand-pipe. The door strongly hinged and fastened. The bottom arranged for torch lighting and extin-

guishing, so as not to leave an opening for the
unrestricted rush of the wind during rough weather.

In places near the sea, or where the lamps are
much exposed, the flaps at the bottom for torch
lighting must be specially arranged, or they will be
blown up and the lights extinguished on every
stormy night.

The shape, such as to admit of the use of flatted
glass entirely, or of some simple and inexpensive
form of bent glass.

Above all it is important that every part of the
lantern be made to gauge, so that, in the event of
portions of it becoming defective, it may be readily
replaced without the expense of removing it from
the column.

The application of such general principles as
these must result in the production of lanterns
suitable for the use to which they are to be put,
while their neglect has been most undeniably the
prime agent in the production of those which now
ornament our streets.

Their most readily perceived defects are a too
abundant illumination of the sky, and the opposite
effect upon the ground.

The shadows of the ribs and ironwork about
them are such, that, projected over the road, they
are exceedingly annoying to drivers of vehicles,
while, to timid pedestrians, they are looked upon as

hiding places for thieves. Indeed, as a general rule, a lamp-post may be said to be an excellent strategic point for garotters and other like-minded individuals. The advancing pedestrian is prevented by the light from the lantern itself from seeing, till he is close to it, whether or not there is anyone standing against the lamp-post. This is especially the case in country districts, and it is precisely in these places where the remedy is most easily applied.

In town it is equally as important, for police purposes, that the fronts of houses should be lighted up to the roof, as it is that the pavement itself should be so. Hence comes the argument in favour of glass tops to lanterns. But this does not apply in the case of hundreds of thousands of lanterns fixed along roads where there are no houses or where the houses are low. A reflecting top fixed in these would utilise those rays of light, in number and intensity not inconsiderable, which are now projected up into the sky, for the exclusive benefit of owls, bats, and mosquitoes, who perhaps do not appreciate the kindness, and would properly direct them to the service of those for whose use they are intended.

Even in streets a proper, and at the same time easy, adjustment of the height of the burner will enable lighting authorities to obtain all the advantages of reflecting tops, without at all interfering with the illumination of the fronts of houses.

WESTM INSTER PATTERN WITH PORCELAIN REFLECTING TOP

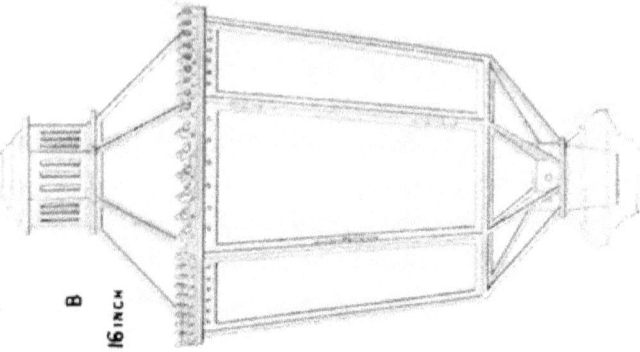

B
16 INCH

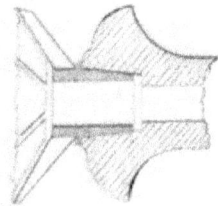

ORDINARY 14 INCH

A

LANTERN WITH PORCELAIN REFLECTING TOP

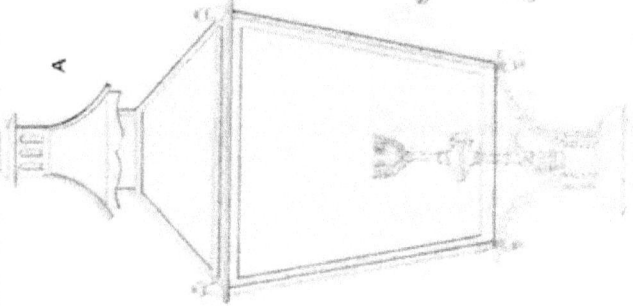

On Plate 6 are the designs for two lanterns suitable for street lighting, which will be found in a marked degree to fulfil the necessary requirements.

Lantern (A) is made in three parts, viz:—the knob and ventilator, the reflector top, and the body. Although this arrangement is carried out more with a view to packing than anything else, yet it is found to offer advantages which, in the opinion of some, warrant its adoption in those cases where it is not necessary for the purposes of transport.

Lantern (B) is rather a modification of the hexagonal lantern than altogether a new pattern, and its principal features are, that it does not require any lamp head, and is besides so arranged, that the bottom need not be broken for the purposes of lighting and extinguishing.

It is also fitted with an improved bolt, which holds the lighting flap very firmly, and prevents the wind from blowing it up in any weather.

It drops firmly into the top of the column, from which however it can easily be removed if necessary.

The top is reflecting and, as well as the knob, is made of porcelain.

APPENDIX.

PUBLIC LIGHTING BY THE DOUBLE TAP SYSTEM.

In the Paper read by Mr. Thos. H. Methven, on page 26, mention is made of this system, the adoption of which was at one time thought by some few admirers of it to be a certain remedy against dissatisfaction either on the part of lighting authorities or Gas Companies.

Although not originally invented by the late Mr. Samuel Hughes, yet undoubtedly to him it owes that amount of popularity, not however large, which it obtained. His faith in it as a means whereby an equitable adjustment of the difficulties between the contracting parties in the matter of public lighting could be arrived at was very great, and he made several trials in various towns, without, however, obtaining any definite result in its favour. It was finally submitted to the Lighting Committee of the Corporation of Folkestone about the beginning of the year 1867, as a "proper regulating apparatus"

within the meaning of that section of the Act relating
to public lighting, obtained during the previous
session by the local Gas Company.

The Folkestone Gas Company, advised by their
Engineer, Mr. Barlow, strongly opposed the idea of
Mr. Hughes, and submitted " Sugg's Self-acting
Governor " as an instrument more nearly approaching
to that " proper regulating apparatus " mentioned in
their Act.

At a conference between the two parties, it was
resolved that Messrs. Hughes and Barlow should
jointly institute certain experiments, with a view
finally to determine which of the two systems proposed was the best, and the report made to the Gas
Company by Mr. Barlow is, by his kind permission,
here given. This is deemed sufficiently conclusive,
as it entirely establishes the superiority of the " Self-
acting Lamp Governor " over the " Double Tap."

MR. BARLOW'S REPORT

ON

DOUBLE TAPS AND SELF-ACTING

REGULATORS

FOR

PUBLIC LIGHTS AT FOLKESTONE.

42, *Parliament Street, London, S.W., April* 4, 1867.

To the Directors of the Folkestone Gas Company.

GENTLEMEN,

In conformity with the instructions I received at the conference you had with the Lighting Committee of the Corporation of Folkestone, on December 11th, I have been in communication with Mr. Hughes, and have made a series of experiments on double taps adjusted by himself, and on Sugg's self-acting regulators, for the purpose of ascertaining whether either of these instruments was a " proper regulating apparatus," within the **meaning** of the 41st section of your Act, for securing uniformity of consumption between metered and unmetered lamps, and have now to report the result.

I stated to Mr. Hughes, at the outset of our conference, that the higher portion of the town, above the Town Hall, was supplied by an independent main from that supplying the lower portion, and that the variations in the pressure during the lighting hours would be less on the high-level main than on the low-level main, so that it might be expedient to try experiments on each level. At a conference I had with him on December 21st, he handed me a paper explanatory of the mode in which he purposed proceeding in adjusting the first series of twelve double taps, which he proposed to fix in the Sandgate Road. I subjoin a copy of this paper, in order that the subsequent proceedings may be more clearly understood :—

"FOLKESTONE GAS.

"*Experiments on the Double Tap.*

" 1. I propose that a group of twelve lamps be selected (comprised within as small a circle as possible).

" 2. That a dry meter be attached to one of these.

" 3. That a double tap and burner be then attached to this **metered lamp and to** the eleven unmetered.

" 4. That the whole twelve lamps be then adjusted, by means of an experimental meter and the governing tap, to a consumption of 2¼ feet **an** hour during the usual day pressure.

" 5. That a stamped meter be then **affixed to each lamp, and the consumption of** gas ascertained during —— nights.

" 6. That, should any discrepancy appear in the registration of the meters, the latter shall be again tested by an inspector under the Sale **of** Gas Acts.

" 7. That, either simultaneously or immediately afterwards, a precisely similar experiment be made with twelve street governors, adjusted to consume 3 feet an hour.

" *Dec.* 20, 1866."   **(Signed)**   " S. HUGHES.

I at once informed Mr. Hughes that I did not purpose interfering in any manner with the adjustment of the double taps, but left him to adopt any plan which he thought would best contribute to the success of the experiment, and I confirmed this in writing a few days subsequently.

Mr. Hughes **commenced the** adjustment **of the twelve double taps,** in the Sandgate Road, on Wednesday, February 6th, **but was** compelled by the boisterous state of the weather to suspend his operations for that day, and the adjustments **were not** completed till the evening of the 8th. In consequence, I presume, of the intimation I had given him, that the night pressure in the Sandgate **Road was** but slightly in excess of the day pressure, he abandoned his idea of adjusting the taps to a consumption of 2¼ feet **per hour, and adopted** 3 feet instead; **and for the purpose** of securing, as **nearly as was** possible under this **system of** adjustment, **uniformity of consumption** between the several lamps, he **employed a wet experimental meter,** which was fixed **on** trestles **and carried about** from lamp to lamp, till, by adjusting the gauge-tap, the consumption of each lamp was stated by Mr. Hughes to be 3 feet an hour, when it was secured by soldering **it** firmly in **its** position; and, for the sake of further security, Mr. Hughes attached a wax **seal to** each. A one-light dry meter was then fixed to each post, and the gas conducted to and from it by flexible tubing to the double tap, and the consumption of each lamp noted for twenty-four hours, at the expiration of which the double taps were removed and replaced by Sugg's regulators, and the consumption of each lamp again noted for another twenty-four hours, with the following results.

Series No. 1

Of Experiments on Twelve Double Taps [adjusted in situ] and Twelve Regulators, in Sandgate Road, Folkestone, with a Meter attached to every Lamp, and a second Meter to Lamp No. 6.

Mr. Hughes's Double Taps.	Consumpt. per hour.	Sugg's Regulators.	Consumpt. per hour.
No. 1	2·54	No. 1	3·05
„ 2	3·46	„ 2	3·25
„ 3	2·96	„ 3	2·66
„ 4	3·12	„ 4	2·58
„ 5	8·70	„ 5	3·91
„ 6 [Metered lamp] . . .		„ 6 [Metered lamp] . .	
„ 7	4·29	„ 7	3·05
„ 8	5·00	„ 8	2·71
„ 9	6·04	„ 9	3·21
„ 10	3·37	„ 10	3·46
„ 11	2·58	„ 11	3·00
„ 12	2·50	„ 12*	
Average . .	**4·06**	**Average . .**	**3·03**

No. 6, Double Tap, with two meters interposed . .	3·21	No. 6, Regulator, with two meters interposed . .	3·00
Average excess per hour of consumpt. of unmetered over metered lamp . .	·85	Average excess per hour of consumpt. of unmetered over metered lamp . .	0·03
Average excess per cent. of consumpt. of unmetered over metered lamp . .	26·48	Average excess per cent. of consumpt. of unmetered over metered lamp . .	1·00

* The Burner of this Regulator was so broken in carriage that it could not be used.

It was apparent to the eye, from the very commencement of this experiment, that this system of adjustment was a perfect failure, and that instead of securing uniformity of consumption between metered and unmetered lamps, it did not even secure uniformity between the unmetered lamps themselves. At first I was inclined to attribute some of the irregularities to imperfections in the meters, and I therefore tested several of them, but found them sufficiently accurate for the purposes of the experiment, though the excessive consumption of No. 5 may be partly due to leakage in the flexible tube. The result shows that the Company would have supplied 26½ per cent. more gas than they would have been paid for, had they accepted double taps thus adjusted as proper regulating instruments within the meaning of the 41st section of their Act. With Sugg's regulators the loss would only have been 1 per cent.

On the conclusion of this series of experiments, the double taps and regulators were taken off the lamps and removed to the Company's works, when they were tested by means of a gasholder containing one cubic foot adjusted to the day pressure at the stand pipe of each lamp. The index of this gasholder was so graduated that it could be read off to the 1000th part of a cubic foot, and the consumption per hour was deduced from the quantity consumed in ten minutes. The following table indicates the results obtained in this second series of experiments :—

SERIES No. 2

Of Experiments on the Double Taps and Regulators at Folkestone, removed from the Lamps in the Sandgate Road.

Mr. Hughes's Double Taps.		Consumpt. per hour.	Sugg's Regulators.		Consumpt. per hour.
No.	1	2·628	No.	1	3·270
,,	2	3·433	,,	2	3·180
,,	3	2·576	,,	3	3·132
,,	4	3·192	,,	4	3·076
,,	5	7·332	,,	5	3·192
,,	6 [Metered lamp] .		,,	6 [Metered lamp] .	
,,	7	4·728	,,	7	3·330
,,	8	5·692	,,	8	3·114
,,	9	6·342	,,	9	2·820
,,	10	3·156	,,	10	3·420
,,	11	3·174	,,	11	3·356
,,	12	2·520	,,	12 [was damaged and not fixed] .	
	Average .	4·095		Average . .	3·188
No. 6, Double tap, with a meter interposed, per hour .		2·916	No. 6, Regulator, with a meter interposed . .		3·012
Average excess per hour of consumpt. of unmetered over metered lamp .		1·179	Average excess per hour of unmetered over metered lamp . . .		0·176
Average excess per cent. of consumpt. of unmetered over metered lamp .		40·43	Average excess per cent. of consumpt. of unmetered over metered lamp . . .		5·84

This series is even more unfavourable to the double tap than the first series, inasmuch as it shows that the Company would have supplied 40½ per cent. more gas than they would have been paid for, had they consented to the use of such an instrument, the night pressure being uniform with the day pressures.

Seeing the unsatisfactory working of the system of adjustment by the double tap adopted in the experiments in the Sandgate Road lamps, Mr. Hughes proposed to adjust a second series of taps for twelve lamps in Harbour Street and Tontine Street, by noting the day pressure at the stand pipe of each lamp, and then regulating the gauge-tap so as to obtain a uniform consumption of 3 cubic feet per hour under that pressure, in the manner followed in the second series of the foregoing experiments. These **twelve double taps I tested, in conjunction** with Mr. Hughes, **on March 9th, with the** following results :—

SERIES No. 3

Of Experiments on Twelve Double Taps, to consume 3 feet per hour at the day pressure, in Tontine and Harbour Streets, Folkestone.

	Consumpt. per hour in cubic feet.
No. 1 [Double tap]	2·916
„ 2	3·048
„ 3	2·970
„ 4	2·868
„ 5	3·156
„ 6	2·952
„ 7 [Metered lamp]	
„ 8	3·042
„ 9	3·108
„ 10	3·133
„ 11	3·180
„ 12	3·144
Average	3·047
No. 7, Double tap, with a meter interposed . . .	2·568
Average excess per hour of consumpt. of unmetered over metered lamp	3·479
Average **excess** per **cent. of consumpt. of unmetered** over metered lamp	18·65

In **this** series of experiments, the approach to uniformity **is nearer than** in the previous ones, but still the Company would have supplied **18½ per** cent. more gas than they would have been paid **for, had** they accepted the double taps thus regulated as a proper regulating instrument within the meaning of the Act.

I invited Mr. **Hughes to make any** further experiments he **thought desirable, and offered to** place my apparatus and laboratory **at his disposal, but he saw no** necessity for continuing the experiments. I understand, however, that he made some, to which I am **not a party, with the view of** showing that the light given by **equal volumes of gas** is greater, when the pressure under which it **is consumed is reduced** by the double tap, than when regulators are **employed.** If this be so, the same advantage may be derived by using **larger** burners in the regulators, and reducing the pressure under which the gas is consumed, which can be done as efficiently and with far less trouble in regulators than in double taps.

Upon the whole, I am of opinion that the double taps I have **tested are not proper regulating** instruments within the meaning of the 41st section of your Act, inasmuch as they do not secure uniformity of consumption between metered and unmetered lamps, and I therefore advise you not to consent to their being adopted for the public lamps of Folkestone.

I append copies of the correspondence that has passed between myself and Mr. Hughes relative to this matter, and cannot refrain from testifying to the uniform courtesy and frankness he has exhibited in dealing with the question; but should his report to the Corporation require any further explanation on my part, I may make it the subject of a supplementary report.

I am, Gentlemen, yours faithfully,

THOMAS G. BARLOW.

The Tables on page 68, which have been very carefully prepared by Mr. Medhurst, the Secretary of the Company, and are here given by the kind permission of Mr. Barlow, show very clearly how well founded is the opinion formed of the Lamp Governors by Mr. Barlow after his experiments, that they are such a "proper regulating apparatus" as is contemplated by the Company's Act of Parliament.

The first part of the Table shows conclusively that the position of the Governor, whether "high" or "low,"—*i.e.*, on high or low ground,—does not in any way influence the consumption of the burner attached to it. For example, in the first column the highest consumption is marked "Medium," the next highest is "Low," while the lowest consumption is "Low." In the second column the highest consumption is "High," and the lowest is "High." In the third column the highest consumption is "Low," while the two lowest are "High," and so on through the rest of the columns.

E

The principal cause of variation is due to slight irregularities in lighting and extinguishing.

An analysis of the second Table shows that on the whole, the arrangements, both as regards quantity of Gas supplied and regularity in lighting and extinguishing, have been thoroughly well carried out.

Thus: the supply in the Quarter ending September 30th, 1868, exceeded the estimated contract quantity by very nearly 10 per cent.; in the December Quarter following it was only 1 per cent. in excess; in the March Quarter of 1869 it was $1\frac{1}{4}$ per cent. in excess; in the following June Quarter 9 per cent. in excess. The grand average for the year being only 4 per cent. above the quantity calculated on.

Again, in the September Quarter of 1869, the consumption was 6 per cent. *over* the estimated quantity, and in the December following it was $1\frac{1}{2}$ per cent. *under* it.

In March, 1870, it was again *under* the estimated quantity by 4 per cent., and in the June following 10 per cent. *over* it.

The grand average for the year is only 1 per cent. *over* the quantity required to be supplied.

Further comment is unnecessary. Actual practice has completely established the correctness of the assumption that it is possible, by the means pointed out in this work, to arrive at a satisfactory solution of the vexatious question, How can the public lamps be supplied with Gas in a manner to satisfy all parties?

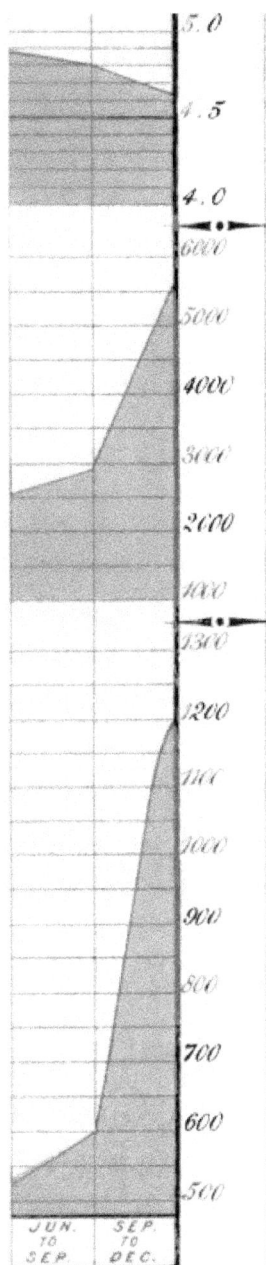

5. 0

This part of the Diagram shews the rate
of consumption per hour of each Lamp,
being the total average consumption of each
Lamp per quarter divided over the number
of hours of burning The irregularities shewn
are therefore from all causes

4 . 5

4 . 0

6000

This part of the Diagram shews the average
consumption of each Lamp The total number
of Lamps in the Town is about 90 and the
number of metered Lamps ranged from
72 in 1866 to 75 in 1870.

5000

4000

3000

2000

1000

1300

This part of the Diagram shews the
number of hours during which the lamps
were burning

1200

1100

This is computed from the Company's
Time Table, but there is one instance
viz. June 2? 1866 in which it is specially
mentioned that it is believed that the number of hours stated is largely in excess of
the actual number. The Lamplighters
are not under the control of the Gas Comp?
Irregularity in lighting will affect the
uppermost compartment of the Diagram

1000

900

800

700

600

500

JUN.
TO
SEP.

SEP.
TO
DEC.

IMPROVED
REGISTERING PRESSURE GAUGE,
With The Drum In The Centre

REGISTER

PRESSURE.	EXHAUST.
To shew 3	2 Vac: & 3 Pressure
5	4 2
6	

List of Prices.

		£	s.	d.
360.	Lamp Governors, with Incorrodible Metal Cone and Burners, complete . . each	0	3	0
361.	Plate 3 (B).—Improved Brass Cocks, "Hawksley Pattern," for ½-in. Standpipes, with Levers, suitable for pole lighting . . . . . each	0	2	0
361a.	Plate 3 (E).—Improved Brass Cocks, Elbow for Bracket Lanterns, with Levers, each	0	2	3
361b.	Ditto ditto, ⅜-in. and ½-in. straight each	0	2	0
362.	Plate 3 (C).—Improved Patent Cocks, with Gun-metal Plug and Patent Metal Vertical Barrel, the Lever working horizontally, with Stop, suitable for pole lighting . . each	0	2	4
363.	Plate 3 (F).—Lighting Torches, in brass, "Sugg's Pattern," with rod, complete, . each	0	12	0
364.	Cleaning Brushes . . per dozen	0	6	0
365.	Plate 3.—Flap Bottoms for Lanterns (two quarters), and strong Glass Flap for 14-in. Lanterns . per dozen	0	15	0
367.	Plate 3 (A).—14-in. Lanterns, glazed and painted, with 16-oz. glass, with Flap Bottoms, complete . . . each	0	10	6
367a.	Ditto ditto, fitted with Porcelain Knob and Porcelain Reflecting Tops . .	0	14	6

£ s. d.

367b. Ditto ditto, with Pegs **and Knob-nuts** for
fixing to lamp-head . . . 0 15 6

368. Plate 6 (A).—14-in. Lanterns (Sugg's
Pattern), constructed in 3 pieces, with
screwed wires **and** metal nuts for putting
together, **and** pegs and knob-nuts for
attaching to lamp-heads. Flap bottom.
Porcelain Knobs and White Porcelain
Reflecting Tops, complete, and painted
good hard colours. . . . each 0 17 6

NOTE.—*This is made to take in pieces for the
convenience of packing.*

369. Plate 6 (B).—Improved "Westminster
Pattern" Hexagonal Lantern, with
Porcelain Reflecting Tops and Conical
Bottom, with Flap Bottom and Improved
Bolt for lighting and extinguishing by
the torch, and gun-metal fitting for
dropping into top of column . . .

*These lanterns require no lamp-head, as they
fix to the top of the column, and are complete in themselves.*

ARTICLES FOR REPAIRS OF LAMP GOVERNORS.

Leathers in **Tin Boxes,** per doz. . . . 0 6 0
Card Washers, in Boxes for Gasways, per gross . 0 1 3
Paper Washers, for Shields, per **gross** . . 0 0 9
Paper Washers, 3 . . . ,, . . . 0 6 9
Solution for Leathers, in Tin Bottles, including
Bottle, per lb. 0 2 6

List of Prices

OF

LAMP METERS AND BOXES.

		£	s.	d.

370. Lamp Meter upon the compensating principle, with extra sized measuring drum, 1 foot in diameter, and of the capacity of $\frac{1}{5}$ of a cubic foot, arranged to work with a friction not exceeding half a tenth of an inch of water, gun metal, differential Index registering up to 10,000 cubic feet—the dial showing single feet, tens, hundreds and thousands, the index box arranged so that it can be filled with oil up to the Dial plate   .   .   .   .   .   each **3 10 0**

371. Cast Iron Box for enclosing Meter, fitted with lid and india-rubber washer, screws and cotters, caps and linings to fit Meter, made sound into box, with plates and back nuts, and external cover roughed so as to be safe for passengers or traffic. The whole complete, ready for fixing   .   each **2 7 0**

372. Syphon Syringe, for extracting waste water from Meters (see page 49).   .   each **0 16 0**

List of Prices

OF

APPARATUS FOR TESTING LAMP GOVERNORS.

		£	s.	d.
5b.	**Plate 2** (C) —One King's Pressure Gauge, showing four inches of pressure, Japanned complete	3	0	0
5.	Plate 2 (D).—One ditto ditto, showing 1½ inches of pressure, sub-divided into 100ths of an inch, Japanned	2	10	0
3.	**Plate 2** (A).—One Experimental Meter, with **measuring drum,** which equals per revolution the rate of 5 cubic feet per hour, into which number the dial is divided, each foot being sub-divided into 100ths of a foot, Japanned	4	4	0
	Plate 2 (F).—Two double **Dry** Governors, each, 33s.	3	6	0
324.—	One Minute Stop-clock, 4-inch dial, divided into 50 divisions, to suit 5 cubic feet per hour calculations; Lever Escapement, strikes each minute, fixed on brass pillar, with weight	6	10	0
	Plate 2.—A **Float** of Lights, fitted with 3 half-inch column cocks, including 2 half-inch main cocks, 1 three-eighth main cock, half-inch tees caps, and plugs, &c , being **the** fittings as shown	2	0	0

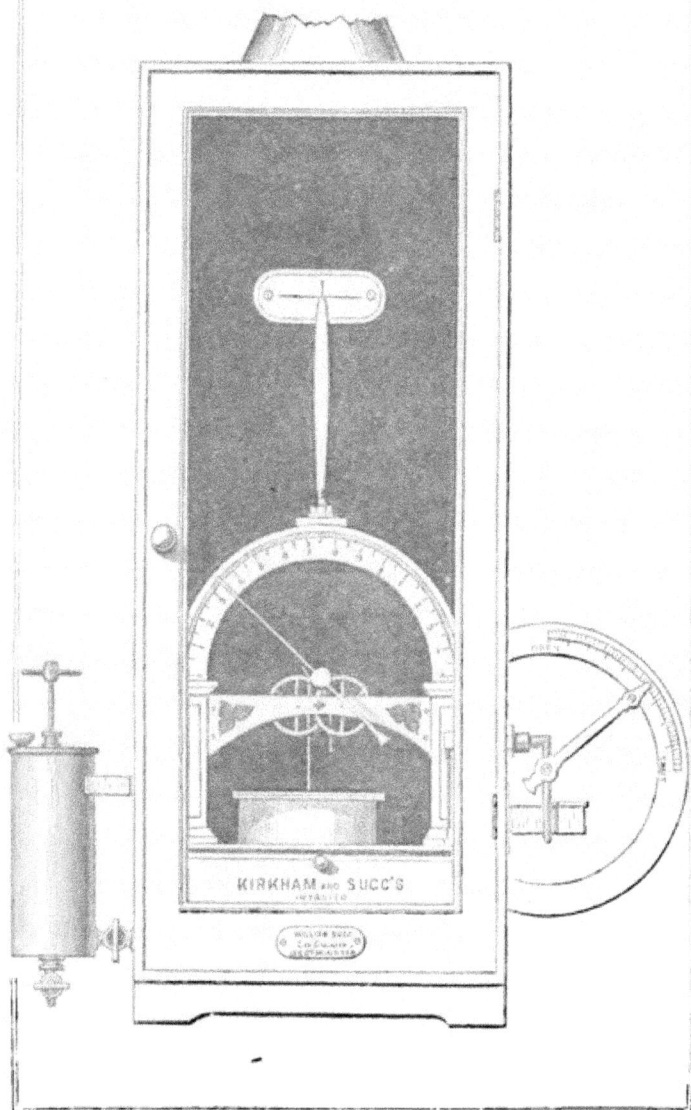

LOWE'S JET PHOTOMETER.

KIRKHAM and SUCC'S

LONDON BURNERS
FITTED
WITH VARIOUS SHADES

THESE SHADES
CAN BE HAD

EITHER PLAIN
OR ORNAMENTED

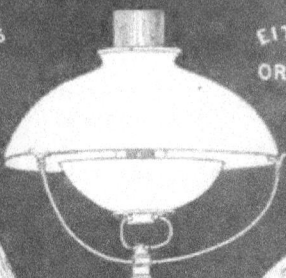

THE ITALIENNE.

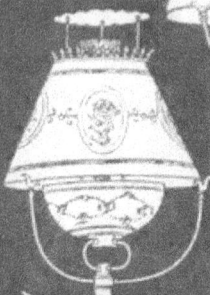

THE WESTMINSTER.

THE VIENNESE.

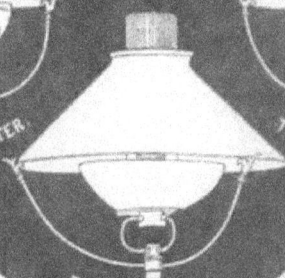

THE FRANKFORT.

THE ROMAN
FLAT FLAME.

THE BRUXELLES
PAPER SHADE

THE PARISIENNE

BATSWING FISHTAIL
REGULATING

BATSWING HOLLOW TOP
REGULATING

These Burners give a greater
amount of light for the quantity
of Gas consumed than any other
Burner yet made.
Vide Gas Referees' Report, March 1869.

www.ingramcontent.com/pod-product-compliance
Lightning Source LLC
Chambersburg PA
CBHW021953190326
41519CB00009B/1241